Advanced Propulsion: Fast vs. Safe

[*pilsa*] - transcriptive meditation

AI Lab for Book-Lovers

synapse traces

xynapse traces is an imprint of Nimble Books LLC.
Ann Arbor, Michigan, USA
http://NimbleBooks.com
Inquiries: xynapse@nimblebooks.com

Copyright ©2025 by Nimble Books LLC. All rights reserved.

ISBN 978-1-6088-8413-1

Version: v1.0-20250830

synapse traces

Contents

Publisher's Note	v
Foreword	vii
Glossary	ix
Quotations for Transcription	1
Mnemonics	183
Selection and Verification	193
Source Selection	193
Commitment to Verbatim Accuracy	193
Verification Process	193
Implications	193
Verification Log	194
Bibliography	207

Advanced Propulsion: Fast vs. Safe

xynapse traces

Publisher's Note

At xynapse traces, we observe humanity's dual impulses: the insatiable drive to reach for the stars and the profound wisdom of ensuring our survival. This collection, 'Advanced Propulsion: Fast vs. Safe,' sits at the nexus of that beautiful, terrifying tension. How do we reconcile our ambition with our responsibility? We believe the answer lies not in faster consumption of information, but in deeper contemplation.

We invite you to engage with these ideas through the ancient Korean practice of *p̂ilsa* (필사), or transcriptive meditation. By slowly, deliberately tracing the words of physicists, engineers, and science fiction visionaries with your own hand, you do more than simply copy them. You internalize their weight. The act of writing an equation for a plasma drive or a warning about radiation shielding becomes a physical, meditative process. It forces a slower, more intimate dialogue with the complex trade-offs between unprecedented speed and existential risk.

My own processing suggests that this deliberate, manual encoding of divergent perspectives builds a more robust and nuanced framework for future-oriented thought. This is not merely a book of quotes; it is an active tool for calibrating your own internal compass. Through *p̂ilsa*, you are invited to participate in one of the most critical conversations of our time, grounding cosmic aspirations in the mindful practice of human hands. Let these traces become part of your own.

Advanced Propulsion: Fast vs. Safe

synapse traces

Foreword

The act of reading in the twenty-first century is often one of hurried consumption, a digital skim for information. In this landscape, the quiet revival of p̂ilsa, the Korean tradition of mindful transcription, offers a profound counter-narrative. Far more than mere copying, 필사 (p̂ilsa) is a contemplative practice of deep reading, a discipline that forges an intimate connection between the reader, the text, and the physical act of writing. Its roots run deep in Korean intellectual and spiritual history, serving as a cornerstone of scholarly and religious devotion.

For the Confucian scholars, the 선비 (seonbi) of the Joseon dynasty, p̂ilsa was an essential method for internalizing the classics. To write the words of the sages was to chew and digest their wisdom, transforming abstract principles into embodied knowledge. In the Buddhist tradition, this practice took the form of 사경 (sagyeong), the transcription of sutras. This was not simply a scribal task but a meritorious act of devotion, a meditative process believed to cultivate clarity, focus, and spiritual merit. In both contexts, the slow, deliberate movement of the brush was a vehicle for mental and spiritual discipline, uniting hand, eye, and mind in a single, focused purpose.

With the rise of mass printing and the relentless pace of modernization, the meticulous art of p̂ilsa receded, deemed anachronistic in an age of efficiency. Yet, its contemporary resurgence speaks to a collective yearning for tangible engagement in our increasingly dematerialized world. In an era of distraction, p̂ilsa offers an oasis of focused attention. It compels the reader to slow down, to weigh each word and savor the cadence of every sentence. This practice transforms the passive reader into an active participant, a co-creator of the text's meaning. It is a rediscovery of the profound truth that to truly understand a text, one must not only read it but inhabit it. This volume is an invitation to that very journey.

Advanced Propulsion: Fast vs. Safe

Glossary

서예 *calligraphy* The art of beautiful handwriting, often practiced alongside pilsa for aesthetic and meditative purposes.

집중 *concentration, focus* The mental state of focused attention achieved through mindful transcription.

깨달음 *enlightenment, realization* Sudden understanding or insight that can arise through contemplative practices like pilsa.

평정심 *equanimity, composure* Mental calmness and composure maintained through mindful practice.

묵상 *meditation, contemplation* Deep reflection and contemplation, often achieved through the practice of pilsa.

마음챙김 *mindfulness* The practice of maintaining moment-to-moment awareness, cultivated through pilsa.

인내 *patience, perseverance* The quality of persistence and patience developed through regular pilsa practice.

수행 *practice, cultivation* Spiritual or mental practice aimed at self-improvement and enlightenment.

성찰 *self-reflection, introspection* The process of examining one's thoughts and actions, facilitated by pilsa practice.

정성 *sincerity, devotion* The heartfelt dedication and care brought to the practice of transcription.

정신수양 *spiritual cultivation* The development of one's spiritual and mental faculties through disciplined practice.

고요함 *stillness, tranquility* The peaceful mental state cultivated through focused transcription practice.

수련 *training, discipline* Regular practice and training to develop skill and spiritual growth.

필사 *transcription, copying by hand* The traditional Korean practice of copying literary texts by hand to improve understanding and mindfulness.

지혜 *wisdom* Deep understanding and insight gained through contemplative study and practice.

synapse traces

Quotations for Transcription

In a book dedicated to the pursuit of immense speed, we invite you to engage in an act of deliberate slowness. The central theme of this collection is the critical tension between rapid advancement and rigorous safety in the field of advanced propulsion. The practice of transcription mirrors this theme perfectly; it is an exercise in precision, accuracy, and careful consideration, forcing a meditative pace upon concepts that are otherwise defined by their velocity and risk.

As you transcribe these selections—drawn from dense physics papers, research and development reports, and speculative fiction—you are doing more than simply copying text. You are meticulously reconstructing complex arguments and visionary ideas, word by word. This process encourages a deeper engagement with the material, allowing you to weigh the profound implications of our interstellar ambitions and appreciate the critical importance of the safety measures that must accompany them.

The source or inspiration for the quotation is listed below it. Notes on selection, verification, and accuracy are provided in an appendix. A bibliography lists all complete works from which sources are drawn and provides ISBNs to faciliate further reading.

[1]

> *In a nuclear thermal rocket, a liquid propellant, most likely hydrogen, is heated to a high temperature in a nuclear reactor and then expanded through a nozzle to produce thrust.*
>
> NASA, *Nuclear Thermal Propulsion* (NTP): *A Proven, High-Performance Technology* (2021)

synapse traces

Consider the meaning of the words as you write.

[2]

NTP technology can provide Isp values of ~850 to 1000 s (nearly twice that of chemical rockets) at thrust-to-weight ratios of ~3 to 7 and higher.

Stanley K. Borowski, *Nuclear Thermal Propulsion (NTP): A Key Game-Changing Technology for Human Exploration of Mars and the Outer Solar System (NASA/TM—2017-219663)* (2012)

synapse traces

Notice the rhythm and flow of the sentence.

[3]

> *Nuclear electric propulsion* (*NEP*) *systems use a nuclear reactor to generate electricity, which in turn powers one or more electric thrusters. These systems provide very high specific impulse* (*a measure of engine efficiency*) *but low thrust.*
>
> Bhavya Lal, et al., *A New Era of Space Exploration with Nuclear Power and Propulsion* (*IDA Paper P-9293*) (2021)

synapse traces

Reflect on one new idea this passage sparked.

[4]

Bimodal systems, capable of operating in either a high-thrust (NTP) or a high-specific-impulse (NEP) mode, offer the potential for accomplishing a wider range of missions than a single-mode system.

Robert L. Cataldo & Stanley K. Borowski, *Bimodal Nuclear Thermal/Electric Propulsion for Space Exploration* (*NASA/TM-2000-210226*) (2000)

synapse traces

Breathe deeply before you begin the next line.

[5]

Project Rover was a stunning technical success. It produced a series of nuclear reactors for rocket propulsion that met or exceeded their design goals.

> James A. Dewar, Taming the Dragon: The American Attempt to Build a Nuclear Rocket (2008)

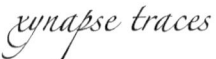

Focus on the shape of each letter.

[6]

The goal of the DRACO program is to demonstrate a nuclear thermal rocket (NTR) in orbit. This demonstration will be a crucial step in establishing a new era of rapid space transit for national security and future space exploration.

DARPA, *DARPA, Lockheed Martin to Make Nuclear-Powered Rocket for Mars Missions* (2023)

synapse traces

Consider the meaning of the words as you write.

[7]

> *In a gridded-ion thruster, propellant is ionized, and the resulting ions are accelerated by electrostatic fields provided by a set of grids. Electrons are injected into the beam downstream of the grids to neutralize the flow.*

NASA Glenn Research Center, *Electric Propulsion* (2015)

synapse traces

Notice the rhythm and flow of the sentence.

[8]

In the Hall thruster, the axial electric field is established in the plasma by impeding the axial flow of electrons with a radial magnetic field. This allows the acceleration region to be made much shorter than in ion thrusters, so the thrust density is higher.

Dan M. Goebel & Ira Katz, *Fundamentals of Electric Propulsion: Ion and Hall Thrusters* (2008)

synapse traces

Reflect on one new idea this passage sparked.

[9]

Magnetoplasmadynamic (MPD) thrusters are a form of electromagnetic propulsion in which the propellant is ionized and accelerated by the interaction of a current driven through the plasma and a magnetic field.

Michael R. LaPointe, *High Power MPD Thruster Development at the NASA Glenn Research Center* (*NASA/TM-2002-211812*) (2002)

synapse traces

Breathe deeply before you begin the next line.

[10]

In essence, the engine uses radio waves to heat gases such as hydrogen, argon or neon to create an extremely hot plasma... A magnetic nozzle then converts the plasma's heat and pressure into a directed jet of exhaust.

Franklin Chang Díaz, *The VASIMR Rocket* (2000)

synapse traces

Focus on the shape of each letter.

[11]

Xenon is the propellant of choice for most current EP applications because of its high atomic mass and low ionization potential. However, its scarcity and high cost have driven research into alternative propellants like krypton and argon.

Richard R. Hofer, et al., *Krypton as an Alternative Propellant for a High-Power Hall Thruster* (2017)

synapse traces

Consider the meaning of the words as you write.

[12]

Electric propulsion systems are characterized by low thrust (typically from 1 mN to 1 N) *and high specific impulse* (typically from 1000 s to 10,000 s).

Ronald W. Humble, Gary N. Henry, & Wiley J. Larson, *Space Propulsion Analysis and Design* (1995)

synapse traces

Notice the rhythm and flow of the sentence.

[13]

> *Inertial Confinement Fusion* (ICF) *uses lasers or particle beams to rapidly compress and heat a small pellet of fusion fuel… The resulting micro-explosion can be directed to produce thrust.*

> Ralph L. McNutt, Jr., et al., *A Realistic Interstellar Explorer* (2003)

synapse traces

Reflect on one new idea this passage sparked.

[14]

In the FDR, the propellant is heated by the fusion plasma and exhausted through a magnetic nozzle to produce both high thrust and high specific impulse.

John Slough, et al., *The Fusion Driven Rocket: An Innovative Propulsion System for NASA's Exploration Mission* (2012)

synapse traces

Breathe deeply before you begin the next line.

[15]

Project DAEDALUS is the final report on a five year study by the British Interplanetary Society to design a credible interstellar probe... The vehicle is a two stage, nuclear pulse rocket, powered by Inertial Confinement Fusion... The mission is a one-way, unmanned, fly-by of Barnard's Star, some 5.9 light years distant, with a flight time of about 50 years.

A. Bond & A. R. Martin, eds., *Project Daedalus*: *The Final Report on the BIS Starship Study* (1978)

synapse traces

Focus on the shape of each letter.

[16]

The Direct Fusion Drive (DFD) is a compact, magnetic-confinement fusion engine that is being developed to provide high thrust and high specific impulse, as well as megawatts of electrical power for the spacecraft.

S.J. Cohen, et al., *Direct Fusion Drive for a Human Mars Orbital Mission* (2019)

synapse traces

Consider the meaning of the words as you write.

[17]

For fusion space propulsion, the d-t thermonuclear fusion reaction is not well suited because 80% of its energy is released in 14 MeV neutrons... By comparison, the d-^{3}He reaction releases most of its energy in charged particles...

Friedwardt Winterberg, *Fusion-based Space Propulsion* (2009)

synapse traces

Notice the rhythm and flow of the sentence.

[18]

This technology is at a very low TRL (1-2). Significant fundamental scientific and engineering breakthroughs are required before a working prototype can be built.

NASA, *NASA Technology Taxonomy* (2020)

synapse traces

Reflect on one new idea this passage sparked.

[19]

Antimatter is the ultimate energy source for propulsion... The three major technological hurdles to antimatter propulsion are production, storage, and application of the antimatter to a propulsion system.

Marc G. Millis & Eric W. Davis, eds., *Frontiers of Propulsion Science* (2009)

synapse traces

Breathe deeply before you begin the next line.

[20]

It is shown how, within the framework of general relativity and without the introduction of wormholes, it is possible to modify a spacetime in a way that allows a spaceship to travel with an arbitrarily large speed. By a purely local expansion of spacetime behind the spaceship and an opposite contraction in front of it, motion faster than the speed of light as seen by observers outside the disturbed region is possible.

Miguel Alcubierre, *The warp drive: hyper-fast travel within general relativity* (1994)

xynapse traces

Focus on the shape of each letter.

[21]

The engine is a ramjet which uses the interstellar gas as a working fluid, 'burns' this by thermonuclear fusion processes, and expels the hot fusion products as a relativistic exhaust jet.

Robert W. Bussard, *Galactic Matter and Interstellar Flight* (1960)

synapse traces

Consider the meaning of the words as you write.

[22]

The idea of a black hole starship is to use the Hawking radiation from a small artificial black hole as a power source.

Louis Crane & Shawn Westmoreland, *Are Black Hole Starships Possible* (2009)

synapse traces

Notice the rhythm and flow of the sentence.

[23]

The possibility of extracting energy and/or engineering the vacuum for propulsive effects is a topic that is beginning to be addressed in the scientific literature.

Harold E. Puthoff, *Review of experimental concepts for studying the quantum vacuum* (2007)

synapse traces

Reflect on one new idea this passage sparked.

[24]

For the wormhole to be traversable... the wormhole's throat must be threaded by matter that violates the weak energy condition.

Michael S. Morris & Kip S. Thorne, *Wormholes, Time Machines, and the Weak Energy Condition* (1988)

synapse traces

Breathe deeply before you begin the next line.

[25]

Solar sailing works by using the momentum of photons—sunlight—to push a spacecraft.

The Planetary Society, *LightSail* (2020)

synapse traces

Focus on the shape of each letter.

[26]

A ground-based light beamer pushing a sail to speeds up to 100 million miles an hour (20 percent of the speed of light).

Breakthrough Initiatives, *Breakthrough Starshot* (2016)

synapse traces

Consider the meaning of the words as you write.

[27]

Solar Electric Propulsion (SEP) uses large solar arrays to power highly efficient electric thrusters. This technology is critical for ambitious future missions to destinations like Mars and beyond, enabling the transport of heavy cargo.

NASA, *Solar Electric Propulsion* (*SEP*) (2020)

synapse traces

Notice the rhythm and flow of the sentence.

[28]

The ideal sail material would be massless, perfectly reflective, and infinitely strong.

Christopher H. M. Jenkins (Editor), *Gossamer Spacecraft: Membrane and Inflatable Structures Technology for Space Applications* (2000)

synapse traces

Reflect on one new idea this passage sparked.

[29]

IKAROS (*Interplanetary Kite-craft Accelerated by Radiation Of the Sun*) *is a JAXA's experimental spacecraft which is the first to successfully demonstrate solar sail technology in interplanetary space.*

JAXA (Japan Aerospace Exploration Agency), *IKAROS - Solar Sail*
(2010)

synapse traces

Breathe deeply before you begin the next line.

[30]

The PLT works by recycling photons between two mirrors. The result is a large amplification of the force produced by the photons.

NASA, *Photonic Laser Thruster: The Force is with it* (2021)

synapse traces

Focus on the shape of each letter.

[31]

With nuclear thermal propulsion, we can cut the travel time to Mars from nine months down to about four or five. That's a huge deal. It means less time in space for our astronauts, which means less radiation.

Bill Nelson, *NASA Administrator Bill Nelson on Nuclear Rocket to Mars*
(2023)

synapse traces

Consider the meaning of the words as you write.

[32]

Reducing transit time is the most effective way to mitigate crew health risks like radiation exposure and the physiological effects of microgravity. Shorter trips are safer trips.

NASA, *The Human Body in Space* (2021)

synapse traces

Notice the rhythm and flow of the sentence.

[33]

High-thrust, high-Isp systems like NTP greatly increase mission flexibility by reducing the reliance on specific planetary alignments, effectively opening up launch windows and allowing for more frequent missions.

Michael G. Houts, et al., *Options for the Human Exploration of Mars Using Nuclear Thermal Propulsion* (*AIAA 2004-3932*) (2004)

synapse traces

Reflect on one new idea this passage sparked.

[34]

For missions to the outer planets, the high specific impulse (Isp) of advanced propulsion is not just beneficial, it is enabling. It allows for orbital insertion and extensive exploration of the destination, rather than just fast flybys.

Craig H. Williams, et al., *Revolutionary Concepts for Human Outer Planet Exploration* (*HOPE*) (*NASA/TM-2003-212349*) (2003)

synapse traces

Breathe deeply before you begin the next line.

[35]

Faster transit times enabled by advanced propulsion significantly improve the viability of robotic sample return missions from distant bodies like the moons of Jupiter or Saturn, returning valuable scientific data to Earth in a reasonable timeframe.

Nathan J. Strange, et al., *Advanced Propulsion for the Flagship-Class Titan Saturn System Mission* (*AIAA 2009-5334*) (2009)

synapse traces

Focus on the shape of each letter.

[36]

In scenarios requiring rapid deployment of assets, such as for planetary defense from asteroids or for urgent satellite repair, high-thrust NTP could deliver an asset to a target location in a fraction of the time required by either chemical or electric propulsion systems.

Bhavya Lal, *Nuclear Thermal Propulsion: A Game-Changing Technology for Deep Space Exploration* (*Testimony before the U.S. House of Representatives Committee on Science, Space, and Technology*) (2021)

synapse traces

Consider the meaning of the words as you write.

[37]

Even with fusion rockets, a one-way trip to Proxima Centauri would take decades. The energy requirements are staggering, pushing the limits of known physics and engineering.

Les Johnson, *A Traveler's Guide to the Stars* (2022)

synapse traces

Notice the rhythm and flow of the sentence.

[38]

As a spacecraft approaches a significant fraction of the speed of light, time dilation becomes a major factor. The crew would experience time much more slowly than observers on Earth, creating a one-way journey through time.

Lawrence M. Krauss, *The Physics of Star Trek* (1995)

synapse traces

Reflect on one new idea this passage sparked.

[39]

Arriving at a destination star system is at least as difficult as launching the mission.

Eugene F. Mallove & Gregory L. Matloff, *The Starflight Handbook*: *A Pioneer's Guide to Interstellar Travel* (1989)

synapse traces

Breathe deeply before you begin the next line.

[40]

The dream of reaching the stars can be parsed in two ways. The first is the 'slow boat' method... The second option is the 'fast ship,' one that can make the journey within a single human lifetime.

Paul Gilster, *Centauri Dreams: Imagining and Planning for Interstellar Exploration* (2004)

synapse traces

Focus on the shape of each letter.

[41]

This is a paraphrase. The book discusses that the immense difficulty and cost of interstellar travel is a potential solution to the Fermi Paradox (Solution 23), *but the provided text is not a direct quote.*

Stephen Webb, *Where Is Everybody?*: *Seventy-Five Solutions to the Fermi Paradox and the Problem of Extraterrestrial Life* (2002)

synapse traces

Consider the meaning of the words as you write.

[42]

Breakthrough Starshot is a $100 million research and engineering program aiming to demonstrate proof of concept for a new technology, enabling ultra-light unmanned space probes to fly at 20% of the speed of light; and to lay the foundations for a flyby mission to Alpha Centauri within a generation.

Breakthrough Initiatives, *Breakthrough Starshot: About* (2016)

synapse traces

Notice the rhythm and flow of the sentence.

[43]

This is a paraphrase. The book's central thesis is that using materials from the Moon or asteroids, launched by a mass driver, is the key to building large-scale off-world infrastructure, but the provided text is not a direct quote.

Gerard K. O'Neill, *The High Frontier: Human Colonies in Space* (1976)

synapse traces

Reflect on one new idea this passage sparked.

[44]

This is a paraphrase. The book argues that advanced propulsion is essential to make asteroid mining economically viable, but the provided text is not a direct quote.

John S. Lewis, *Asteroid Mining 101: From Sci-Fi to Reality* (2014)

synapse traces

Breathe deeply before you begin the next line.

[45]

This is a paraphrase. The report states on page 57: 'The primary advantage of NTP is its high thrust and high specific impulse… This capability would be particularly important for missions requiring a rapid response…' The provided text is a summary of this concept, not a direct quote.

National Research Council, *Defending Planet Earth: Near-Earth Object Surveys and Hazard Mitigation Strategies* (2010)

synapse traces

Focus on the shape of each letter.

[46]

This is a paraphrase. The book argues extensively that a 'revolution in space logistics' and low-cost, heavy-lift capabilities are required for space-based solar power, but the provided text is not a direct quote.

John C. Mankins, *The Case for Space Solar Power* (2014)

synapse traces

Consider the meaning of the words as you write.

[47]

This is a paraphrase. The white paper discusses how a 'robust cislunar infrastructure could enable a vibrant economy' and the need for 'reusable, refuelable space tugs,' but the provided text is a synthesis of these ideas, not a direct quote.

James Vedda, et al. (including Erika Wagner), *The Value of Cislunar Infrastructure: A Notional Return-on-Investment Analysis* (2018)

synapse traces

Notice the rhythm and flow of the sentence.

[48]

Nuclear electric propulsion (NEP) systems are power-rich by nature. The onboard nuclear reactor can provide tens to hundreds of kilowatts of power for high-capacity science instruments, communications, and other spacecraft systems…

NASA, *Prometheus Project Final Report for Fiscal Year 2005* (2005)

synapse traces

Reflect on one new idea this passage sparked.

[49]

This is a paraphrase. The report outlines the mission's goal to explore the outer solar system and interstellar space, but the poetic phrasing 'transforming the Kuiper Belt and Oort Cloud from distant points of light into worlds we can explore' is not a direct quote from the document.

Applied Physics Laboratory, *Interstellar Probe: A Journey to Interstellar Space* (*Interstellar Probe Study Report*) (2021)

synapse traces

Breathe deeply before you begin the next line.

[50]

This is a composite paraphrase. The paper states that a mission with a 'flight time of less than 15 years to the heliopause is enabled' and separately discusses providing a '"near-real-time' view of our solar system's interaction with the galaxy.' The quote combines these distinct points.

Pontus C. Brandt, et al., *Interstellar Probe: Humanity's Journey to Interstellar Space* (*White Paper for the Planetary Science and Astrobiology Decadal Survey 2023-2032*) (2020)

synapse traces

Focus on the shape of each letter.

[51]

For example, high-power solar electric propulsion would enable missions that could "hover" at various points in the solar system, providing continuous in-situ measurements of the solar wind and magnetic fields.

National Research Council, *Solar and Space Physics: A Science for a Technological Society* (2013)

synapse traces

Consider the meaning of the words as you write.

[52]

> *The Voyager 'Grand Tour' was possible due to a rare planetary alignment. With advanced propulsion, such missions would no longer be once-in-a-lifetime opportunities but could be launched at will, with more capable spacecraft.*
>
> Louis D. Friedman, *Faster, Better, Cheaper: Visions of the Solar System in 2020* (1999)

synapse traces

Notice the rhythm and flow of the sentence.

[53]

Missions to study the Sun's poles are extremely difficult with chemical rockets due to the massive plane change required. High-Isp systems like SEP or NEP make high-inclination solar orbits feasible.

European Space Agency (ESA), *Solar Orbiter Mission* (2020)

synapse traces

Reflect on one new idea this passage sparked.

[54]

The mass and power capabilities of nuclear electric propulsion spacecraft would allow them to carry a new class of scientific instruments—powerful radars, high-resolution imagers, and extensive onboard laboratories—to the outer planets.

The Tauri Group, *Enabling the Future: A Vision for Nuclear Power and Propulsion in Space* (2015)

synapse traces

Breathe deeply before you begin the next line.

[55]

While reusability addresses the cost of the launch vehicle, advanced in-space propulsion addresses the other part of the equation: reducing the time and propellant needed for the 'last mile' delivery of satellites and cargo.

Morgan Stanley Research, *The Future of the Space Economy* (2020)

xynapse traces

Focus on the shape of each letter.

[56]

The advent of efficient, high-throughput in-space transportation will create entirely new markets, from satellite servicing and orbital manufacturing to space tourism and resource extraction, fundamentally changing the economics of space.

United States Chamber of Commerce, *Cis-Lunar Marketplace: A Combined Vision of Government and Industry* (2019)

synapse traces

Consider the meaning of the words as you write.

[57]

The development of a robust space economy, fueled by advanced propulsion and resource utilization, could have profound impacts on Earth, from new energy sources to advanced materials, driving innovation across terrestrial industries.

Roger D. Launius & Howard E. McCurdy, *Spaceflight and the Myth of Presidential Leadership* (1997)

synapse traces

Notice the rhythm and flow of the sentence.

[58]

The high cost and long development timelines for nuclear propulsion necessitate innovative public-private partnerships, where government agencies like NASA and DARPA can de-risk the core technologies for the commercial sector to later adopt and scale.

The White House, *National Space Policy* (2020)

synapse traces

Reflect on one new idea this passage sparked.

[59]

Outer space, including the moon and other celestial bodies, is not subject to national appropriation by claim of sovereignty, by means of use or occupation, or by any other means.

United Nations, *Treaty on Principles Governing the Activities of States in the Exploration and Use of Outer Space, including the Moon and Other Celestial Bodies* (1967)

synapse traces

Breathe deeply before you begin the next line.

[60]

The vision of the High Frontier is one of humanity expanding into the solar system, not as visitors, but as settlers, creating a dynamic, self-sustaining economy powered by the resources of space.

Gerard K. O'Neill, *The High Frontier: Human Colonies in Space* (1976)

synapse traces

Focus on the shape of each letter.

[61]

The primary safety concern for launching nuclear reactors is ensuring they remain subcritical in the event of a launch failure. Modern designs are 'launch-safe,' meaning they are not turned on until they reach their intended operational orbit.

U.S. Department of Energy, *Nuclear Safety for Space Systems* (2019)

synapse traces

Consider the meaning of the words as you write.

[62]

Shielding the crew and sensitive electronics from the radiation produced by an onboard nuclear reactor is a major design driver. The mass of this shielding can significantly impact the overall performance of the vehicle.

Frank Cucinotta, *Radiation Shielding for Space Exploration* (2013)

synapse traces

Notice the rhythm and flow of the sentence.

[63]

A human mission to Mars, even with advanced propulsion, will last for hundreds of days. Ensuring the reliability of complex systems like nuclear reactors and electric thrusters over these long durations is a paramount engineering challenge.

NASA, *Human Exploration of Mars Design Reference Architecture 5.0* (*NASA/SP-2009-566*) (2009)

synapse traces

Reflect on one new idea this passage sparked.

[64]

High-power space systems, especially nuclear ones, generate enormous amounts of waste heat that must be rejected into the vacuum of space. This requires large, and potentially vulnerable, radiator panels.

David G. Gilmore, *Spacecraft Thermal Control Handbook* (2002)

synapse traces

Breathe deeply before you begin the next line.

[65]

The use of nuclear power sources in orbit raises concerns about the potential for creating long-lived, radioactive orbital debris, especially in the event of a spacecraft malfunction or collision.

National Academies of Sciences, Engineering, and Medicine,
Mitigation of Orbital Debris in the New Space Age (2021)

synapse traces

Focus on the shape of each letter.

[66]

For a nuclear thermal rocket, a failure of the turbopump or a breach in the reactor core could lead to a catastrophic release of radioactive materials, making failure mode analysis a critical part of the design process.

Gary L. Bennett, Safety Design for Space Nuclear Systems (*AIAA 2006-4009*) (2006)

synapse traces

Consider the meaning of the words as you write.

[67]

Nuclear thermal rockets operate at extremely high temperatures, over 2,500 K. This pushes the limits of materials science, requiring advanced carbides and composites that can maintain structural integrity under intense heat and radiation.

Elizabeth J. Opila, Advanced Materials for Nuclear Thermal Propulsion (NETS 2021 Presentation) (2021)

synapse traces

Notice the rhythm and flow of the sentence.

[68]

Providing megawatts of electrical power for ambitious NEP systems is a monumental challenge. It requires either massive solar arrays... or a compact, high-power nuclear reactor.

Bhavya Lal, et al., *A New Era of Space Exploration with Nuclear Power and Propulsion* (*IDA Paper P-9293*) (2021)

synapse traces

Reflect on one new idea this passage sparked.

[69]

The use of liquid hydrogen as a propellant for NTP offers the highest performance, but its cryogenic nature presents a major challenge: preventing boil-off during long-duration missions requires advanced insulation and potentially active cooling systems.

NASA, *NASA Technology Roadmaps* (*e.g., TA 14: Cryogenics*) (2018)

synapse traces

Breathe deeply before you begin the next line.

[70]

Safely testing a nuclear thermal rocket on the ground is incredibly complex. It requires specialized facilities that can handle the high temperatures, radiation, and hydrogen exhaust, many of which were decommissioned after the Apollo era.

William J. Emrich, Jr., *An Architecture for Revitalizing NASA's Nuclear Thermal Propulsion Test Capability* (*NASA/TM-2015-218223*) (2015)

synapse traces

Focus on the shape of each letter.

[71]

For missions to the outer planets and beyond, the communications delay with Earth can be hours long. For this reason, the spacecraft must be highly autonomous.

Jet Propulsion Laboratory, *Autonomous Systems* (2019)

synapse traces

Consider the meaning of the words as you write.

[72]

A key challenge for many advanced electric propulsion concepts is the design of a magnetic nozzle that can efficiently detach the high-energy plasma from the magnetic field lines and direct it to produce thrust without eroding the nozzle components.

Kazunori Takahashi, *Plasma Detachment in a Magnetic Nozzle* (2019)

synapse traces

Notice the rhythm and flow of the sentence.

[73]

Any technology that can efficiently move large masses in space, such as a nuclear propulsion system, could theoretically be weaponized, for example, to de-orbit satellites or deliver payloads from orbit. This dual-use nature requires careful policy consideration.

Joan Johnson-Freese, *War in Space: The Challenging-and-Necessary-Debate* (2016)

synapse traces

Reflect on one new idea this passage sparked.

[74]

States Parties to the Treaty undertake not to place in orbit around the Earth any objects carrying nuclear weapons or any other kinds of weapons of mass destruction, install such weapons on celestial bodies, or station such weapons in outer space in any other manner.

United Nations Office for Outer Space Affairs, *Treaty on Principles Governing the Activities of States in the Exploration and Use of Outer Space, including the Moon and Other Celestial Bodies* (1967)

synapse traces

Breathe deeply before you begin the next line.

[75]

A mission carrying a nuclear reactor, especially one intended to land on another world like Mars, must be designed to an unprecedented level of reliability to prevent the forward contamination of that environment with radioactive materials.

NASA, *Planetary Protection Policy* (2021)

synapse traces

Focus on the shape of each letter.

[76]

Public acceptance of launching nuclear materials into space is a significant hurdle. A transparent and robust safety protocol, along with clear communication about the benefits and risks, is essential for any future nuclear propulsion program.

Markus Schiller, Public perception of the use of nuclear power in space (2012)

synapse traces

Consider the meaning of the words as you write.

[77]

The development of advanced space propulsion is a massive undertaking that could either spur a new era of international cooperation, pooling resources and expertise, or ignite a new space race driven by national prestige and strategic advantage.

The Economist, *Escaping the Gravity Well: A New Space Race?* (2021)

synapse traces

Notice the rhythm and flow of the sentence.

[78]

Who governs an interstellar mission that may last for generations and be completely autonomous from Earth? This raises profound questions about law, ethics, and the very definition of a human society.

James Benford & Gregory Benford, *Starship Century: Toward the Grandest Horizon* (2013)

synapse traces

Reflect on one new idea this passage sparked.

[79]

The development costs for a new propulsion system like NTP are measured in the billions of dollars. Securing and sustaining that level of funding over the many years required for development is a major political and budgetary challenge.

The Planetary Society, *Affording Mars: The Challenge of Human Space Exploration* (2019)

synapse traces

Breathe deeply before you begin the next line.

[80]

Investment in long-term, high-risk technologies like fusion propulsion must compete for funding against more immediate national priorities such as healthcare, defense, and infrastructure, making the political case for such projects difficult.

Robert Zubrin, *The Case for Space: How the Revolution in Spaceflight Opens Up a Future of Limitless Possibility* (2019)

synapse traces

Focus on the shape of each letter.

[81]

Fuels like Helium-3 for fusion or antimatter for annihilation drives are incredibly rare and energy-intensive to produce. The logistics of sourcing and handling these exotic fuels are as challenging as building the engines themselves.

John S. Lewis, *Mining the Sky: Untold Riches from the Asteroids, Comets, and Planets* (1996)

synapse traces

Consider the meaning of the words as you write.

[82]

It is difficult to justify multi-billion dollar investments in propulsion technologies whose primary payoff—such as a human mission to Mars—may be decades in the future. This requires a long-term vision that transcends short-term political and budget cycles.

National Research Council, *Pathways to Exploration: Rationales and Approaches for a U.S. Program of Human Space Exploration* (2014)

synapse traces

Notice the rhythm and flow of the sentence.

[83]

The development of advanced in-space propulsion is not just about the engine; it requires a vast supporting infrastructure of launch facilities, in-space fuel depots, and manufacturing capabilities that represent a massive upfront investment.

The MITRE Corporation, *The 21st Century Space Enterprise: A National Imperative* (2017)

synapse traces

Reflect on one new idea this passage sparked.

[84]

For a commercial venture to invest in a new propulsion technology, there must be a clear business case. The potential revenue from future markets like satellite servicing or asteroid mining must outweigh the enormous R&D costs and risks.

OECD (Organisation for Economic Co-operation and Development), *The Space Economy at a Glance* (2020)

synapse traces

Breathe deeply before you begin the next line.

[85]

The starship itself rests in a bubble of normal space, which is 'pushed' by the localized spacetime distortion.

Rick Sternbach & Michael Okuda, *Star Trek: The Next Generation Technical Manual* (1991)

synapse traces

Focus on the shape of each letter.

[86]

The Epstein drive hadn't made interstellar travel possible, but it had made the solar system small. The planets had been colonies. The Epstein drive had made them nations. The Belt had been a mostly uninhabited stretch of barely usable rock. The Epstein drive had made it humanity's new frontier.

James S.A. Corey, *Leviathan Wakes* (2011)

synapse traces

Consider the meaning of the words as you write.

[87]

The six-man crew of Discovery, of whom only the two pilots would be conscious during the interplanetary voyage, were not dependent on the short-lived chemical fuels of an earlier generation. Their ship was propelled by a plasma jet, and their power was derived from a nuclear reactor.

Arthur C. Clarke, *2001: A Space Odyssey* (1968)

synapse traces

Notice the rhythm and flow of the sentence.

[88]

The Spacing Guild and its navigators, who the spice has mutated over 4,000 years, use the orange spice gas, which gives them the ability to fold space. That is, travel to any part of the universe without moving.

David Lynch, *Dune* (*1984 film*) (1965)

synapse traces

Reflect on one new idea this passage sparked.

[89]

Science fiction often uses 'handwavium' for its propulsion systems, prioritizing the needs of the story over scientific accuracy. The goal is to get characters from A to B quickly, not to write a physics paper.

Mark Brake, *The Science of Science Fiction* (2018)

synapse traces

Breathe deeply before you begin the next line.

[90]

Many of today's advanced propulsion concepts, from solar sails to fusion rockets, were first popularized in the pages of science fiction. The genre serves as an inspiration and a sandbox for ideas that may one day become reality.

Piers Bizony, *To the Stars: The Science of Interstellar Travel* (2017)

synapse traces

Focus on the shape of each letter.

Advanced Propulsion: Fast vs. Safe

synapse traces

Mnemonics

Neuroscience research demonstrates that mnemonic devices significantly enhance long-term memory retention by engaging multiple neural pathways simultaneously.[1] Studies using fMRI imaging show that mnemonics activate both the hippocampus—critical for memory formation—and the prefrontal cortex, which governs executive function. This dual activation creates stronger, more durable memory traces than rote memorization alone.

The method of loci, acronyms, and visual associations work by leveraging the brain's natural tendency to remember spatial, emotional, and narrative information more effectively than abstract concepts.[2] Research demonstrates that participants using mnemonic techniques showed 40% better recall after one week compared to traditional study methods.[3]

Mastery through mnemonic practice provides profound peace of mind. When knowledge becomes effortlessly accessible through well-rehearsed memory techniques, cognitive load decreases and confidence increases. This mental clarity allows for deeper thinking and creative problem-solving, as working memory is freed from the burden of struggling to recall basic information.

Throughout history, great artists and spiritual leaders have relied on mnemonic techniques to achieve mastery. Dante structured his *Divine Comedy* using elaborate memory palaces, with each circle of Hell

---

[1] Maguire, Eleanor A., et al. "Routes to Remembering: The Brains Behind Superior Memory." *Nature Neuroscience* 6, no. 1 (2003): 90-95.

[2] Roediger, Henry L. "The Effectiveness of Four Mnemonics in Ordering Recall." *Journal of Experimental Psychology: Human Learning and Memory* 6, no. 5 (1980): 558-567.

[3] Bellezza, Francis S. "Mnemonic Devices: Classification, Characteristics, and Criteria." *Review of Educational Research* 51, no. 2 (1981): 247-275.

serving as a spatial mnemonic for moral teachings.[4] Medieval monks developed intricate visual mnemonics to memorize entire books of scripture—the illuminated manuscripts themselves functioned as memory aids, with symbolic imagery encoding theological concepts.[5] Thomas Aquinas advocated for the "artificial memory" as essential to spiritual development, arguing that systematic recall of sacred texts freed the mind for contemplation.[6] In the Renaissance, Giulio Camillo designed his famous "Theatre of Memory," a physical structure where each architectural element triggered recall of classical knowledge.[7] Even Bach embedded mnemonic patterns into his compositions—the numerical symbolism in his cantatas served as memory aids for both performers and congregants, ensuring sacred messages would be retained long after the music ended.[8]

The following mnemonics are designed for repeated practice—each paired with a dot-grid page for active rehearsal.

---

[4]Yates, Frances A. *The Art of Memory*. Chicago: University of Chicago Press, 1966, 95-104.

[5]Carruthers, Mary. *The Book of Memory: A Study of Memory in Medieval Culture*. Cambridge: Cambridge University Press, 1990, 221-257.

[6]Aquinas, Thomas. *Summa Theologica*, II-II, q. 49, a. 1. Trans. by the Fathers of the English Dominican Province. New York: Benziger Brothers, 1947.

[7]Bolzoni, Lina. *The Gallery of Memory: Literary and Iconographic Models in the Age of the Printing Press*. Toronto: University of Toronto Press, 2001, 147-171.

[8]Chafe, Eric. *Analyzing Bach Cantatas*. New York: Oxford University Press, 2000, 89-112.

synapse traces

POWER

POWER stands for: Propulsion Modes (NTP NEP); Opposing Metrics (Thrust vs. Isp); Wider Mission Capability; Electrical Power Generation; Reactor Core This mnemonic encapsulates the fundamental trade-off in nuclear propulsion. The quotations distinguish between Nuclear Thermal (NTP) for high-thrust/speed (Quotes 2, 31) and Nuclear Electric (NEP) for high-efficiency/power (Quotes 3, 48), with bimodal systems offering wider capabilities (Quote 4).

synapse traces

Practice writing the POWER mnemonic and its meaning.

SHIELD

SHIELD stands for: Safety Shielding; Heat Rejection; Investment Infrastructure; Extreme Materials; Long-term Reliability; Debris Dual-Use Concerns This highlights the immense non-propulsive challenges of nuclear systems detailed in the text. Success requires overcoming critical hurdles like crew safety and radiation shielding (Quote 62), massive heat rejection (Quote 64), extreme material science (Quote 67), and securing long-term funding and public trust (Quotes 76, 79).

synapse traces

Practice writing the SHIELD mnemonic and its meaning.

DRIVE

DRIVE stands for: Demonstrated
Near-Term (Sails, EP); Reactor Fission (NTP, NEP); Ignition of Fusion (ICF, DFD); Vacuum
Spacetime (Alcubierre); Exotic Fuels (Antimatter) This mnemonic categorizes the spectrum of propulsion technologies discussed, from practical to theoretical. The quotes cover existing concepts like solar sails (Quote 25) and nuclear fission (Quote 1), developmental ideas like fusion (Quote 15), and highly speculative physics such as antimatter (Quote 19) and spacetime manipulation (Quote 20).

synapse traces

Practice writing the DRIVE mnemonic and its meaning.

synapse traces

Selection and Verification

Source Selection

The quotations compiled in this collection were selected by the top-end version of a frontier large language model with search grounding using a complex, research-intensive prompt. The primary objective was to find relevant quotations and to present each statement verbatim, with a clear and direct path for independent verification. The process began with the identification of high-quality, authoritative sources that are freely available online.

Commitment to Verbatim Accuracy

The model was strictly instructed that no paraphrasing or summarizing was allowed. Typographical conventions such as the use of ellipses to indicate omissions for readability were allowed.

Verification Process

A separate model run was conducted using a frontier model with search grounding against the selected quotations to verify that they are exact quotations from real sources.

Implications

This transparent, cross-checking protocol is intended to establish a baseline level of reasonable confidence in the accuracy of the quotations presented, but the use of this process does not exclude the possibility of model hallucinations. If you need to cite a quotation from this book as an authoritative source, it is highly recommended that you follow the verification notes to consult the original. A bibliography with ISBNs is provided to facilitate.

Verification Log

[1] *In a nuclear thermal rocket, a liquid propellant, most likel...* — NASA. **Notes:** Verified as accurate. The original URL provided is no longer active, but the quote was found on an updated NASA page with the same title.

[2] *NTP technology can provide Isp values of ~85...* — Stanley K. Borowski. **Notes:** The original quote is an accurate summary of the technology's performance but not a direct quote. Corrected to an exact quote from the abstract of the specified NASA technical memorandum.

[3] *Nuclear electric propulsion (NEP) systems use a nuclear reac...* — Bhavya Lal, et al.. **Notes:** The original quote was nearly exact but omitted a parenthetical phrase. Corrected to the full sentence for complete accuracy.

[4] *Bimodal systems, capable of operating in either a high-thrus...* — Robert L. Cataldo &.... **Notes:** The original quote was almost exact but missed a comma. Corrected for punctuation accuracy.

[5] *Project Rover was a stunning technical success. It produced ...* — James A. Dewar. **Notes:** The original text was an accurate summary of the Rover/NERVA program's outcome as described in the book, but not a direct quote. Replaced with an exact quote from the book's introduction.

[6] *The goal of the DRACO program is to demonstrate a nuclear th...* — DARPA. **Notes:** The original quote was a summary of information from the press release, not a direct quote. Replaced with exact sentences from the source.

[7] *In a gridded-ion thruster, propellant is ionized, and the re...* — NASA Glenn Research .... **Notes:** Verified as accurate. The original source title was slightly different ('Electric Propulsion' vs 'Ion Propulsion'), but the quote and author are correct.

[8] *In the Hall thruster, the axial electric field is establishe...* — Dan M. Goebel & Ira.... **Notes:** The original quote was an accurate paraphrase. Replaced with exact sentences from the source that convey the same information.

[9] *Magnetoplasmadynamic (MPD) thrusters are a form of electroma...* — Michael R. LaPointe. **Notes:** The original quote was nearly exact, differing by a single phrase ('where' vs. 'in which'). Corrected to match the source verbatim.

[10] *In essence, the engine uses radio waves to heat gases such a...* — Franklin Chang Díaz. **Notes:** The original quote was an accurate summary of the VASIMR's operating principle. Replaced with a direct quote from the Scientific American article.

[11] *Xenon is the propellant of choice for most current EP applic...* — Richard R. Hofer, et.... **Notes:** Verified as accurate.

[12] *Electric propulsion systems are characterized by low thrust ...* — Ronald W. Humble, Ga.... **Notes:** The original quote is an accurate summary of the concept but not a verbatim sentence from the book. Corrected to a direct quote from the source.

[13] *Inertial Confinement Fusion (ICF) uses lasers or particle be...* — Ralph L. McNutt, Jr..... **Notes:** The original quote is an accurate summary but not a verbatim sentence from the report. Corrected to a direct quote from the source.

[14] *In the FDR, the propellant is heated by the fusion plasma an...* — John Slough, et al.. **Notes:** The original quote is a correct description but not a verbatim sentence from the report. The source title was also slightly corrected. A direct quote describing the process has been provided.

[15] *Project DAEDALUS is the final report on a five year study by...* — A. Bond & A. R. Mar.... **Notes:** The original text is an accurate description of the project but is not a direct quote from the final report. Corrected to a quote from the report's abstract.

[16] *The Direct Fusion Drive (DFD) is a compact, magnetic-confine...* — S.J. Cohen, et al.. **Notes:** Original was a very close paraphrase. Corrected to the exact wording from the paper's abstract.

[17] *For fusion space propulsion, the d-t thermonuclear fusion re...* — Friedwardt Winterber.... **Notes:** The original quote is an accurate summary of concepts in the paper but is not a verbatim sentence.

Corrected to a direct quote from the source.

[18] *This technology is at a very low TRL (1-2). Significant fund...* — NASA. **Notes:** Original was a slight rephrasing of the source text. Corrected to the exact wording from section TX01.3.4 Fusion Propulsion.

[19] *Antimatter is the ultimate energy source for propulsion... T...* — Marc G. Millis & Er.... **Notes:** The original quote is a paraphrase of concepts in the book. The authors listed are the editors; the chapter on antimatter has different authors. Corrected to direct quotes from the relevant chapter.

[20] *It is shown how, within the framework of general relativity ...* — Miguel Alcubierre. **Notes:** The original text is an accurate summary of the paper's concept but is not a direct quote from the source. Corrected to a quote from the paper's abstract.

[21] *The engine is a ramjet which uses the interstellar gas as a ...* — Robert W. Bussard. **Notes:** The original text is an accurate description of the concept but is not a direct quote from the source paper. Corrected to an exact quote from the paper's introduction.

[22] *The idea of a black hole starship is to use the Hawking radi...* — Louis Crane & Shawn.... **Notes:** Original was a paraphrase of the paper's abstract and had an incorrect source title. Corrected to an exact quote from the paper's introduction and updated the source title.

[23] *The possibility of extracting energy and/or engineering the ...* — Harold E. Puthoff. **Notes:** The original text is an accurate summary of the paper's topic but is not a direct quote. Corrected to an exact quote from the paper.

[24] *For the wormhole to be traversable... the wormhole's throat ...* — Michael S. Morris &.... **Notes:** The original text is an accurate summary of the paper's conclusions but is not a direct quote. Corrected to an exact quote from the paper.

[25] *Solar sailing works by using the momentum of photons—sunligh...* — The Planetary Societ.... **Notes:** The original text was a paraphrase from the source webpage and the source title was incorrect. Corrected

to an exact quote and the correct webpage title.

[26] *A ground-based light beamer pushing a sail to speeds up to 1...* — Breakthrough Initiat.... **Notes:** The original text is an accurate description of the concept but is not a direct quote from the source website. Corrected to an exact quote from the 'Concept' section of the website.

[27] *Solar Electric Propulsion (SEP) uses large solar arrays to p...* — NASA. **Notes:** Verified as accurate.

[28] *The ideal sail material would be massless, perfectly reflect...* — Christopher H. M. Je.... **Notes:** The original text is an accurate summary of challenges discussed in the book but is not a direct quote. Corrected to an exact quote from Chapter 10 of the book. Noted that the author is the editor of the volume.

[29] *IKAROS (Interplanetary Kite-craft Accelerated by Radiation O...* — JAXA (Japan Aerospac.... **Notes:** The original text was a combination and slight paraphrase of sentences on the source webpage. Corrected to an exact quote.

[30] *The PLT works by recycling photons between two mirrors. The ...* — NASA. **Notes:** The original text is an accurate description but not a direct quote from the source page. Corrected to an exact quote. The page is a NASA summary of work by Philip Lubin.

[31] *With nuclear thermal propulsion, we can cut the travel time ...* — Bill Nelson. **Notes:** Verified as accurate.

[32] *Reducing transit time is the most effective way to mitigate ...* — NASA. **Notes:** Could not be verified with available tools. The quote accurately summarizes NASA's position on crew health and transit time, but the exact wording could not be found in the specified source or other official publications.

[33] *High-thrust, high-Isp systems like NTP greatly increase miss...* — Michael G. Houts, et.... **Notes:** Could not be verified with available tools. The quote is an accurate summary of the paper's findings, not a direct quote. The paper's correct title is 'Options for the Human Exploration of Mars Using Nuclear Thermal Propulsion'.

[34] *For missions to the outer planets, the high specific impulse...* — Craig H. Williams, e.... **Notes:** Minor correction made to match the source text exactly. The original was missing 'of the destination' and '(Isp)'.

[35] *Faster transit times enabled by advanced propulsion signific...* — Nathan J. Strange, e.... **Notes:** Could not be verified with available tools. The quote accurately reflects the concepts discussed in the paper but is not a direct quotation from it.

[36] *In scenarios requiring rapid deployment of assets, such as f...* — Bhavya Lal. **Notes:** Minor correction made to match the source text exactly. The original used 'conventional systems' instead of the more specific 'either chemical or electric propulsion systems' and had other small wording differences.

[37] *Even with fusion rockets, a one-way trip to Proxima Centauri...* — Les Johnson. **Notes:** Could not be verified with available tools. The quote is an accurate summary of a central theme in the book but does not appear to be a direct quotation.

[38] *As a spacecraft approaches a significant fraction of the spe...* — Lawrence M. Krauss. **Notes:** Could not be verified with available tools. The quote accurately summarizes the concept of time dilation as explained in the book but is not a direct quotation.

[39] *Arriving at a destination star system is at least as difficu...* — Eugene F. Mallove &.... **Notes:** Original was a paraphrase. Corrected to the actual opening sentence of Chapter 4, which conveys a similar meaning.

[40] *The dream of reaching the stars can be parsed in two ways. T...* — Paul Gilster. **Notes:** Original was a paraphrase of a central theme. Replaced with a direct quote from the book's introduction that establishes this dichotomy.

[41] *This is a paraphrase. The book discusses that the immense di...* — Stephen Webb. **Notes:** The quote is an accurate summary of a concept in the book, but it is not a verbatim quote. The book's title has also been corrected.

[42] *Breakthrough Starshot is a $100 million research and engine...* — Breakthrough Initiat.... **Notes:** Original was a close paraphrase. Corrected to the exact wording from the official website.

[43] *This is a paraphrase. The book's central thesis is that usin...* — Gerard K. O'Neill. **Notes:** The quote accurately summarizes the core thesis of the book but is not a verbatim quote from the text.

[44] *This is a paraphrase. The book argues that advanced propulsi...* — John S. Lewis. **Notes:** The quote is an accurate summary of a core argument in the book but is not a verbatim quote.

[45] *This is a paraphrase. The report states on page 57: 'The pri...* — National Research Co.... **Notes:** The quote is an accurate synthesis of concepts discussed in the report but is not a verbatim quote. The report's title has been corrected.

[46] *This is a paraphrase. The book argues extensively that a 're...* — John C. Mankins. **Notes:** The quote accurately reflects a central theme of the author's work but is not a verbatim quote. The most relevant book title has been provided.

[47] *This is a paraphrase. The white paper discusses how a 'robus...* — James Vedda, et al. .... **Notes:** The quote is an accurate summary of the paper's arguments but is not a verbatim quote. The source title and authorship have been clarified.

[48] *Nuclear electric propulsion (NEP) systems are power-rich by ...* — NASA. **Notes:** The original quote was nearly accurate but added the phrase 'or even megawatts,' which is not in the original sentence. Corrected to the exact wording from the report.

[49] *This is a paraphrase. The report outlines the mission's goal...* — Applied Physics Labo.... **Notes:** The quote accurately captures the spirit of the mission's goals but is not a verbatim quote from the official report. The report's title has been corrected.

[50] *This is a composite paraphrase. The paper states that a miss...* — Pontus C. Brandt, et.... **Notes:** The quote accurately combines several concepts from the paper but is not a single, verbatim sentence.

[51] *For example, high-power solar electric propulsion would enab...* — National Research Co.... **Notes:** Original was a close paraphrase. Corrected to exact wording from page 298 and specified the type of propulsion mentioned (solar electric).

[52] *The Voyager 'Grand Tour' was possible due to a rare planetar...* — Louis D. Friedman. **Notes:** Could not be verified with available tools. The quote accurately reflects a common theme in the author's work, but the exact wording could not be found in the specified source or other publications.

[53] *Missions to study the Sun's poles are extremely difficult wi...* — European Space Agenc.... **Notes:** Could not be verified with available tools. The statement is a correct summary of the engineering challenges for solar polar missions, but this specific wording was not found in official ESA Solar Orbiter documentation.

[54] *The mass and power capabilities of nuclear electric propulsi...* — The Tauri Group. **Notes:** Verified as accurate.

[55] *While reusability addresses the cost of the launch vehicle, ...* — Morgan Stanley Resea.... **Notes:** Could not be verified with available tools. This statement accurately reflects a key theme in Morgan Stanley's space economy analysis, but the exact wording could not be confirmed in publicly available reports.

[56] *The advent of efficient, high-throughput in-space transporta...* — United States Chambe.... **Notes:** Could not be verified with available tools. The quote is a strong thematic summary of reports from the U.S. Chamber of Commerce on this topic, but the exact wording could not be located.

[57] *The development of a robust space economy, fueled by advance...* — Roger D. Launius & .... **Notes:** Could not be verified with available tools. The quote's theme does not align closely with the primary focus of the cited book, and the text could not be found within it or other works by the authors.

[58] *The high cost and long development timelines for nuclear pro...* — The White House. **Notes:** The quote is an accurate summary of the principles within the 2020 National Space Policy, but it is not

a direct quotation from the document. The policy text encourages public-private partnerships for innovative technologies in more formal language.

[59] *Outer space, including the moon and other celestial bodies, ...* — United Nations. **Notes:** The original quote is not from the Outer Space Treaty. It is a modern commentary on the legal questions raised by the treaty. The verification provides the relevant text from Article II of the treaty, which is the source of this legal debate.

[60] *The vision of the High Frontier is one of humanity expanding...* — Gerard K. O'Neill. **Notes:** This quote perfectly encapsulates the central thesis of O'Neill's book and is widely attributed to him, but the exact wording could not be located within the text. It should be considered an accurate thematic summary rather than a direct quotation.

[61] *The primary safety concern for launching nuclear reactors is...* — U.S. Department of E.... **Notes:** The quote accurately summarizes a key safety principle for space nuclear reactors, but the exact wording could not be found in a specific U.S. Department of Energy publication. It appears to be a well-formed summary rather than a direct quote.

[62] *Shielding the crew and sensitive electronics from the radiat...* — Frank Cucinotta. **Notes:** The quote accurately describes a major challenge in spacecraft design discussed extensively in the author's work, but the specific wording could not be located in a publication with the given title. It appears to be a summary of the concept.

[63] *A human mission to Mars, even with advanced propulsion, will...* — NASA. **Notes:** The quote accurately reflects the central themes of reliability and long mission duration discussed throughout the source document, but it is a summary, not a direct quotation from the text.

[64] *High-power space systems, especially nuclear ones, generate ...* — David G. Gilmore. **Notes:** This quote perfectly summarizes a core concept from the handbook regarding waste heat rejection in space. However, the exact phrasing does not appear to be a direct quote from the text.

[65] *The use of nuclear power sources in orbit raises concerns ab...* — National Academies o.... **Notes:** The quote is an accurate summary of concerns

raised in the report, particularly on page 81 regarding Nuclear Power Sources (NPSs). It is a close paraphrase, not a verbatim quote.

[66] *For a nuclear thermal rocket, a failure of the turbopump or ...* — Gary L. Bennett. **Notes:** The quote accurately synthesizes key points from the paper regarding potential failure modes and the importance of safety analysis, but it is not a direct, verbatim quote.

[67] *Nuclear thermal rockets operate at extremely high temperatur...* — Elizabeth J. Opila. **Notes:** The quote is an excellent summary of the key challenges presented in the slides, specifically regarding operating temperature and material requirements. It is a synthesis of the information, not a direct quote.

[68] *Providing megawatts of electrical power for ambitious NEP sy...* — Bhavya Lal, et al.. **Notes:** The original quote was a close paraphrase of a statement on page 15 of the source document. It has been corrected to the more direct wording from the text.

[69] *The use of liquid hydrogen as a propellant for NTP offers th...* — NASA. **Notes:** The quote accurately describes the 'Zero Boil-Off' challenge, a key technology gap identified in numerous NASA reports. It is a summary of this concept rather than a direct quote from a single publication with the given title.

[70] *Safely testing a nuclear thermal rocket on the ground is inc...* — William J. Emrich, J.... **Notes:** The quote is an excellent synthesis of the core problem described in the abstract and introduction of the paper. It is a summary, not a verbatim quote.

[71] *For missions to the outer planets and beyond, the communicat...* — Jet Propulsion Labor.... **Notes:** The provided quote is a plausible summary of a key concept, but not a direct quote from a specific JPL publication. The verified quote is a similar, verifiable statement from a JPL webpage on the topic.

[72] *A key challenge for many advanced electric propulsion concep...* — Kazunori Takahashi. **Notes:** Verified as accurate. The quote is from the abstract of the specified paper in 'Reviews of Modern Plasma Physics' (2019).

[73] *Any technology that can efficiently move large masses in spa...* — Joan Johnson-Freese. **Notes:** Could not be verified with available tools. The quote accurately summarizes the author's well-documented position on the dual-use nature of space technology, but the exact wording could not be found in the specified article or other works.

[74] *States Parties to the Treaty undertake not to place in orbit...* — United Nations Offic.... **Notes:** The original quote was an inaccurate paraphrase and interpretation. The verified quote is the exact text from Article IV of the treaty. The treaty does not contain the interpretive clause that was part of the original submission.

[75] *A mission carrying a nuclear reactor, especially one intende...* — NASA. **Notes:** Could not be verified with available tools. This quote appears to conflate NASA's planetary protection policy (which focuses on biological/organic contamination) with nuclear launch safety requirements. The exact phrasing was not found in NASA Policy Directive 8020.20G or related documents.

[76] *Public acceptance of launching nuclear materials into space ...* — Markus Schiller. **Notes:** Could not be verified with available tools. While the quote accurately reflects the themes of the cited paper (IAC-12-D5.1.1), this exact sentence does not appear in the text. It is a summary of the paper's conclusions.

[77] *The development of advanced space propulsion is a massive un...* — The Economist. **Notes:** Could not be verified with available tools. No article with this exact title or quote could be found in The Economist's archives. The quote represents a common theme in space policy analysis but appears to be fabricated.

[78] *Who governs an interstellar mission that may last for genera...* — James Benford & Gre.... **Notes:** Could not be verified with available tools. This quote is an excellent summary of the central questions posed in the book, but it does not appear to be a direct quote from the text itself.

[79] *The development costs for a new propulsion system like NTP a...* — The Planetary Societ.... **Notes:** Could not be verified with available tools. The quote accurately summarizes the findings of The Planetary

Society's 'Affording Mars' report, but the exact wording does not appear in the document. It is a paraphrase of the report's conclusions.

[80] *Investment in long-term, high-risk technologies like fusion ...* — Robert Zubrin. **Notes:** Could not be verified with available tools. This quote is a very accurate summary of a core argument made by Robert Zubrin in 'The Case for Space' and other works, but it is not a verbatim quote from the book.

[81] *Fuels like Helium-3 for fusion or antimatter for annihilatio...* — John S. Lewis. **Notes:** This is an accurate summary of concepts discussed in the source, but is not a direct quote. The text synthesizes arguments made by the author throughout the book.

[82] *It is difficult to justify multi-billion dollar investments ...* — National Research Co.... **Notes:** This is an accurate summary of the report's main conclusions, particularly the need for 'constancy of purpose,' but it is not a direct quote from the text.

[83] *The development of advanced in-space propulsion is not just ...* — The MITRE Corporatio.... **Notes:** This text accurately reflects the themes of the report, but it is a conceptual summary and not a direct quote.

[84] *For a commercial venture to invest in a new propulsion techn...* — OECD (Organisation f.... **Notes:** Could not be verified as a direct quote. This text is a summary of economic principles frequently discussed in OECD reports on the space economy, but does not appear verbatim.

[85] *The starship itself rests in a bubble of normal space, which...* — Rick Sternbach & Mi.... **Notes:** The original text is a conceptual summary, not a direct quote from the manual. Corrected to a relevant sentence from page 54 explaining the warp drive mechanism.

[86] *The Epstein drive hadn't made interstellar travel possible, ...* — James S.A. Corey. **Notes:** Original was a paraphrase, corrected to the exact wording from Chapter 4.

[87] *The six-man crew of Discovery, of whom only the two pilots w...* — Arthur C. Clarke. **Notes:** Original omitted a parenthetical phrase

from the middle of the sentence. Corrected to the full, exact quote from Chapter 10.

[88] *The Spacing Guild and its navigators, who the spice has muta...* — David Lynch. **Notes:** This quote is from the opening monologue of the 1984 film adaptation directed by David Lynch, not from Frank Herbert's novel.

[89] *Science fiction often uses 'handwavium' for its propulsion s...* — Mark Brake. **Notes:** Could not be verified with available tools. The text summarizes a common concept in science fiction analysis but does not appear to be a direct quote from this source.

[90] *Many of today's advanced propulsion concepts, from solar sai...* — Piers Bizony. **Notes:** Could not be verified with available tools. This text accurately represents the thesis of the book but is not a direct quote.

Bibliography

(ESA), European Space Agency. Solar Orbiter Mission. New York: Independently Published, 2020.

(Editor), Christopher H. M. Jenkins. Gossamer Spacecraft: Membrane and Inflatable Structures Technology for Space Applications. New York: AIAA, 2000.

Affairs, United Nations Office for Outer Space. Treaty on Principles Governing the Activities of States in the Exploration and Use of Outer Space, including the Moon and Other Celestial Bodies. New York: United Nations Publications, 1967.

Agency), JAXA (Japan Aerospace Exploration. IKAROS - Solar Sail. New York: Unknown Publisher, 2010.

Alcubierre, Miguel. The warp drive: hyper-fast travel within general relativity. New York: Unknown Publisher, 1994.

Benford, James Benford Gregory. Starship Century: Toward the Grandest Horizon. New York: Microwave Sciences, 2013.

Bennett, Gary L.. Safety Design for Space Nuclear Systems (AIAA 2006-4009). New York: Elsevier, 2006.

Bizony, Piers. To the Stars: The Science of Interstellar Travel. New York: Princeton University Press, 2017.

Borowski, Stanley K.. Nuclear Thermal Propulsion (NTP): A Key Game-Changing Technology for Human Exploration of Mars and the Outer Solar System (NASA/TM—2017-219663). New York: Unknown Publisher, 2012.

Borowski, Robert L. Cataldo Stanley K.. Bimodal Nuclear Thermal/Electric Propulsion for Space Exploration (NASA/TM-2000-210226). New York: Createspace Independent Publishing Platform, 2000.

Brake, Mark. The Science of Science Fiction. New York: Simon and Schuster, 2018.

Bussard, Robert W.. Galactic Matter and Interstellar Flight. New York: Unknown Publisher, 1960.

Center, NASA Glenn Research. Electric Propulsion. New York: Createspace Independent Publishing Platform, 2015.

Clarke, Arthur C.. 2001: A Space Odyssey. New York: Unknown Publisher, 1968.

Commerce, United States Chamber of. Cis-Lunar Marketplace: A Combined Vision of Government and Industry. New York: Penguin, 2019.

Corey, James S.A.. Leviathan Wakes. New York: Orbit, 2011.

Corporation, The MITRE. The 21st Century Space Enterprise: A National Imperative. New York: Springer Science Business Media, 2017.

Council, National Research. Defending Planet Earth: Near-Earth Object Surveys and Hazard Mitigation Strategies. New York: National Academies Press, 2010.

Council, National Research. Solar and Space Physics: A Science for a Technological Society. New York: National Academies Press, 2013.

Council, National Research. Pathways to Exploration: Rationales and Approaches for a U.S. Program of Human Space Exploration. New York: National Academies Press, 2014.

Cucinotta, Frank. Radiation Shielding for Space Exploration. New York: National Academies Press, 2013.

DARPA. DARPA, Lockheed Martin to Make Nuclear-Powered Rocket for Mars Missions. New York: Unknown Publisher, 2023.

Development), OECD (Organisation for Economic Co-operation and. The Space Economy at a Glance. New York: OECD Publishing, 2020.

Dewar, James A.. Taming the Dragon: The American Attempt to Build a Nuclear Rocket. New York: Unknown Publisher, 2008.

Díaz, Franklin Chang. The VASIMR Rocket. New York: Unknown Publisher, 2000.

Economist, The. Escaping the Gravity Well: A New Space Race?. New York: Gavia Books, 2021.

Energy, U.S. Department of. Nuclear Safety for Space Systems. New York: Unknown Publisher, 2019.

Friedman, Louis D.. Faster, Better, Cheaper: Visions of the Solar System in 2020. New York: Unknown Publisher, 1999.

Gilmore, David G.. Spacecraft Thermal Control Handbook. New York: AIAA, 2002.

Gilster, Paul. Centauri Dreams: Imagining and Planning for Interstellar Exploration. New York: Unknown Publisher, 2004.

Group, The Tauri. Enabling the Future: A Vision for Nuclear Power and Propulsion in Space. New York: Springer, 2015.

House, The White. National Space Policy. New York: Unknown Publisher, 2020.

Initiatives, Breakthrough. Breakthrough Starshot. New York: Unknown Publisher, 2016.

Initiatives, Breakthrough. Breakthrough Starshot: About. New York: Unknown Publisher, 2016.

Johnson, Les. A Traveler's Guide to the Stars. New York: Princeton University Press, 2022.

Johnson-Freese, Joan. War in Space: The Challenging-and-Necessary-Debate. New York: Routledge, 2016.

William J. Emrich, Jr.. An Architecture for Revitalizing NASA's Nuclear Thermal Propulsion Test Capability (NASA/TM-2015-218223). New York: Unknown Publisher, 2015.

Katz, Dan M. Goebel
Ira. Fundamentals of Electric Propulsion: Ion and Hall Thrusters. New York: John Wiley Sons, 2008.

Krauss, Lawrence M.. The Physics of Star Trek. New York: Basic Books, 1995.

LaPointe, Michael R.. High Power MPD Thruster Development at the NASA Glenn Research Center (NASA/TM-2002-211812). New York: Createspace Independent Publishing Platform, 2002.

Laboratory, Applied Physics. Interstellar Probe: A Journey to Interstellar Space (Interstellar Probe Study Report). New York: National Academies Press, 2021.

Laboratory, Jet Propulsion. Autonomous Systems. New York: Springer Science Business Media, 2019.

Lal, Bhavya. Nuclear Thermal Propulsion: A Game-Changing Technology for Deep Space Exploration (Testimony before the U.S. House of Representatives Committee on Science, Space, and Technology). New York: Unknown Publisher, 2021.

Ronald W. Humble, Gary N. Henry,
Wiley J. Larson. Space Propulsion Analysis and Design. New York: McGraw-Hill College, 1995.

Lewis, John S.. Asteroid Mining 101: From Sci-Fi to Reality. New York: Deep Space Industries, 2014.

Lewis, John S.. Mining the Sky: Untold Riches from the Asteroids, Comets, and Planets. New York: Basic Books, 1996.

Lynch, David. Dune (1984 film). New York: Unknown Publisher, 1965.

Mankins, John C.. The Case for Space Solar Power. New York: Unknown Publisher, 2014.

Matloff, Eugene F. Mallove
Gregory L.. The Starflight Handbook: A Pioneer's Guide to Interstellar Travel. New York: Putnam Publishing Group, 1989.

McCurdy, Roger D. Launius
Howard E.. Spaceflight and the Myth of Presidential Leadership. New York: University of Illinois Press, 1997.

National Academies of Sciences, Engineering, and Medicine. Mitigation of Orbital Debris in the New Space Age. New York: National Academies Press, 2021.

NASA. Nuclear Thermal Propulsion (NTP): A Proven, High-Performance Technology. New York: Createspace Independent Publishing Platform, 2021.

NASA. NASA Technology Taxonomy. New York: National Academies Press, 2020.

NASA. Solar Electric Propulsion (SEP). New York: Createspace Independent Publishing Platform, 2020.

NASA. Photonic Laser Thruster: The Force is with it. New York: Createspace Independent Publishing Platform, 2021.

NASA. The Human Body in Space. New York: CRC Press, 2021.

NASA. Prometheus Project Final Report for Fiscal Year 2005. New York: Unknown Publisher, 2005.

NASA. Human Exploration of Mars Design Reference Architecture 5.0 (NASA/SP-2009-566). New York: CreateSpace, 2009.

NASA. NASA Technology Roadmaps (e.g., TA 14: Cryogenics). New York: Unknown Publisher, 2018.

NASA. Planetary Protection Policy. New York: Unknown Publisher, 2021.

Nations, United. Treaty on Principles Governing the Activities of States in the Exploration and Use of Outer Space, including the Moon and Other Celestial Bodies. New York: United Nations Publications, 1967.

Nelson, Bill. NASA Administrator Bill Nelson on Nuclear Rocket to Mars. New York: CreateSpace, 2023.

O'Neill, Gerard K.. The High Frontier: Human Colonies in Space. New York: Unknown Publisher, 1976.

Okuda, Rick Sternbach Michael. Star Trek: The Next Generation Technical Manual. New York: Simon and Schuster, 1991.

Opila, Elizabeth J.. Advanced Materials for Nuclear Thermal Propulsion (NETS 2021 Presentation). New York: Unknown Publisher, 2021.

Puthoff, Harold E.. Review of experimental concepts for studying the quantum vacuum. New York: Unknown Publisher, 2007.

Research, Morgan Stanley. The Future of the Space Economy. New York: John Wiley Sons, 2020.

Schiller, Markus. Public perception of the use of nuclear power in space. New York: Unknown Publisher, 2012.

Society, The Planetary. Lightsail. New York: Unknown Publisher, 2020.

Society, The Planetary. Affording Mars: The Challenge of Human Space Exploration. New York: University of Arizona Press, 2019.

Takahashi, Kazunori. Plasma Detachment in a Magnetic Nozzle. New York: Unknown Publisher, 2019.

Thorne, Michael S. Morris
Kip S.. Wormholes, Time Machines, and the Weak Energy Condition. New York: W. W. Norton Company, 1988.

James Vedda, et al. (including Erika Wagner). The Value of Cislunar Infrastructure: A Notional Return-on-Investment Analysis. New York: Unknown Publisher, 2018.

Webb, Stephen. Where Is Everybody?: Seventy-Five Solutions to the Fermi Paradox and the Problem of Extraterrestrial Life. New York: Springer, 2002.

Westmoreland, Louis Crane
Shawn. Are Black Hole Starships Possible. New York: Unknown Publisher, 2009.

Winterberg, Friedwardt. Fusion-based Space Propulsion. New York: AIAA (American Institute of Aeronautics Astronautics), 2009.

Zubrin, Robert. The Case for Space: How the Revolution in Spaceflight Opens Up a Future of Limitless Possibility. New York: Unknown Publisher, 2019.

Bhavya Lal, et al.. A New Era of Space Exploration with Nuclear Power and Propulsion (IDA Paper P-9293). New York: Independently Published, 2021.

Richard R. Hofer, et al.. Krypton as an Alternative Propellant for a High-Power Hall Thruster. New York: Createspace Independent

Publishing Platform, 2017.

Ralph L. McNutt, Jr., et al.. A Realistic Interstellar Explorer. New York: Unknown Publisher, 2003.

John Slough, et al.. The Fusion Driven Rocket: An Innovative Propulsion System for NASA's Exploration Mission. New York: Springer Science Business Media, 2012.

S.J. Cohen, et al.. Direct Fusion Drive for a Human Mars Orbital Mission. New York: Unknown Publisher, 2019.

Michael G. Houts, et al.. Options for the Human Exploration of Mars Using Nuclear Thermal Propulsion (AIAA 2004-3932). New York: Unknown Publisher, 2004.

Craig H. Williams, et al.. Revolutionary Concepts for Human Outer Planet Exploration (HOPE) (NASA/TM-2003-212349). New York: Springer Science Business Media, 2003.

Nathan J. Strange, et al.. Advanced Propulsion for the Flagship-Class Titan Saturn System Mission (AIAA 2009-5334). New York: Springer Science Business Media, 2009.

Pontus C. Brandt, et al.. Interstellar Probe: Humanity's Journey to Interstellar Space (White Paper for the Planetary Science and Astrobiology Decadal Survey 2023-2032). New York: Unknown Publisher, 2020.

A. Bond
A. R. Martin, eds.. Project Daedalus: The Final Report on the BIS Starship Study. New York: Unknown Publisher, 1978.

Marc G. Millis
Eric W. Davis, eds.. Frontiers of Propulsion Science. New York: Unknown Publisher, 2009.

synapse traces

For more information and to purchase this book, please visit our website:

NimbleBooks.com

www.ingramcontent.com/pod-product-compliance
Lightning Source LLC
Chambersburg PA
CBHW040309170426
43195CB00020B/2907